KARTHIKEYAN S
ASHWIN R

DEVELOPMENT OF A 3D- PRINTED HUMANOID ROBOT

KARTHIKEYAN S
ASHWIN R

DEVELOPMENT OF A 3D-PRINTED HUMANOID ROBOT

NEXT-GEN HUMAN ASSISTANCE USING 3D-PRINTED HUMANOID ROBOTS

LAP LAMBERT Academic Publishing

Imprint
Any brand names and product names mentioned in this book are subject to trademark, brand or patent protection and are trademarks or registered trademarks of their respective holders. The use of brand names, product names, common names, trade names, product descriptions etc. even without a particular marking in this work is in no way to be construed to mean that such names may be regarded as unrestricted in respect of trademark and brand protection legislation and could thus be used by anyone.

Cover image: www.ingimage.com

Publisher:
LAP LAMBERT Academic Publishing
is a trademark of
Dodo Books Indian Ocean Ltd. and OmniScriptum S.R.L publishing group

120 High Road, East Finchley, London, N2 9ED, United Kingdom
Str. Armeneasca 28/1, office 1, Chisinau MD-2012, Republic of Moldova, Europe
Managing Directors: Ieva Konstantinova, Victoria Ursu
info@omniscriptum.com

Printed at: see last page
ISBN: 978-620-4-20953-1

NEXT-GEN HUMAN ASSISTANCE USING 3D-PRINTED HUMANOID ROBOTS

ABSTRACT

The main objective behind developing this robot is for the human assistance using artificial intelligence. Humanoid robot with dual manipulators and dexterous hands shows great significance in domestic, medical and service applications. They can provide companion, operation, manipulation, material handling and many other services to human beings. However, current humanoid robots are either too expansive or too clumsy. There is a trade-off between robot flexibility and costs. It is a huge obstacle blocking the road of humanoid robot toward our daily life. Machine vision is one of the most important tasks for the interaction between the robot and the environment, because it provides more information about the elements that exist there which makes interactions easier.

TABLE OF CONTENTS

LIST OF FIGURES

TITLE

CHAPTER 1

INTRODUCTION

1.1 NEED FOR THE STUDY

The project uses 3D printing combined with readily available components to make the robot affordable and easily manufacturable. The robot parts were printed with PLA-filament on a desktop 3D printer, it is the most common material used on desktop 3D printers with the right material properties for their application. Desktop 3D printing was also used as it's the most accessible and affordable manufacturing technology worldwide.

The goal of project is to provide human assistance using a humanoid robot because we are promised by film, fiction and television that humanoids will cook for us, clean for us, become our best friends, teach our children, and even fall in love with us. Humanoid robots have human-like design and are able to mimic human motions So human assistance by a Humanoid robot will keep us more comfortable and more interactive to us, the robot can be used in numerous practical applications that along with general assistance can attempt to solve the root of the problem. This robot can be used to provide assistance in places like helping in teaching sign language with a human teacher.

Neural networks, roughly simulate the working of Human brain to make machines learn from examples. AI is intelligence demonstrated by machines, in contrast to natural Intelligence. The term Artificial intelligence is applied when machine mimics "cognitive" functions that human associate with other human minds such as learning and problem solving.

Artificial Intelligence deals with creation of intelligent system. Intelligent systems are those systems which possess intelligence like humans. With the help of computers and availability of large enough data sets. Image Recognition plays a vital role dealing with artificial intelligence. Our brains make vision seem easy. It doesn't take any effort for humans to tell apart a lion and a jaguar, read a sign, or recognize a human's face. But these are actually hard problems to solve with a computer, they only seem easy because our brains are incredibly good at understanding images.

Pattern Recognition in machine learning focus on the recognition of patterns and regularities in data. although it is in some cases considered to be nearly synonymous with machine learning. Pattern recognition systems are in many cases trained from labeled "training" data (supervised learning), but when no labeled data are available other algorithms can be used to discover previously unknown patterns (unsupervised learning).

Speech Recognition is the inter disciplinary sub-field of computational linguistics that develops methodologies and technologies that enables the recognition and translation of spoken languages into text by computers.

CHAPTER 2

REVIEW OF LITERATURE

2.1 INTRODUCTION

The goal of this project is not to replace human sign language translators but to merely provide support when they are not available. Once the designs are optimized, the robot can be used in numerous practical applications that along with general assistance can attempt to solve the root of the problem. The root problem is that sign language courses are sparse resulting in a shortage of the necessary number of translators. This robot can be used to help teach sign language with a human teacher thereby expanding the capacity of these classes.

A deaf person who needs to appear in court, A deaf person following a lesson in a classroom somewhere. These are all circumstances where a deaf person needs a sign language interpreter, but where often such an interpreter is not readily available. This is where a low-cost option, like Humanoid robot arm can offer a solution. Technologies like 3d printing, artificial intelligence comes into play where with technology gets enhanced. Not only with sign language translation human assistance are also being plugged-in to our project. Image recognition provides various uses including people tagging, gaming, price comparison, identifying tv shows, and augmented reality. And speech recognition is AI Assistant That Knows Who You Are, and it works with various application including object detection and image recognition.

2.2 SYSTEM REVIEW

2.2.1 Iain Werry, Kerstin Dautenhahn, Bernard Ogden, - published a paper titled as "Can Social Interaction Skills Be Taught by a Social Agent? The Role of a Robotic Mediator in Autism Therapy". Increasingly socially intelligent agents (software or robotic) are used in education, rehabilitation and therapy. This paper discusses the role of inter-active, mobile robots as social mediators in the particular domain of autism therapy. This research is part of the project AURORA that studies how mobile robots can be used to teach children with autism basic interaction skills that are important in social interactions among humans. Results from a particular series of trials involving pairs of two children and a mobile robot are described. The results show that the scenario with pairs of children and a robot creates a very interesting social context which gives rise to a variety of different social and non-social interaction patterns, demonstrating the specific problems but also abilities of children with autism in social interactions. Future work will include a closer analysis of international structure in human-human and robot-human interaction. We outline a particular framework that we are investigating.

2.2.2 S. Hashimolo, S. Narita, H. Kasahara, A. Takanishi, S. Sugano, K. Shirai, T. Kobayashi, H. Takanobu, T. Kurata, K. Fujiwara, T. Matsuno, T. Kawasaki, K. Hoashi, -published a paper titled as "Humanoid Robot - Development of an Information Assistant Robot Hadaly". At the Advanced Research Institute for Science and Engineering, Waseda University, the Humanoid Project has been in operation since 1992 with the objective of developing an anthropomorphic robot that shares psychological and physical space with humans [l]. An in- formation assistant robot Hadaly (Fig. l), was developed in 1995 with an emphasis on communications between humans and a robot. Some works on information assistant robot have been reported. Matsui et al. [2] proposed an in-house information robot, ancl Kojima et al. [3] proposed a re- receptionist robot that utilizes a mobile robot on the market. In this study, the authors feature an attempt to find the configurations and functions that are required for a humanoid to "assist with information" in intellectual op- erations. "Hadaly" is the name of a robot that appears in a novel written by Villiers de I'lsle-Adam, a 19th century French novelist, titled "L'EVE FUTURE." In this paper the outline of Hadaly is described along with the contents of the four subsystems

2.2.3 David Feil-Seifer, Maja J Matari'c, -published a paper titled as "Human-Robot Interaction". Human-robot interaction (HRI) is the interdisciplinary study of interaction dynamics between humans and robots. Researchers and practitioners specializing in HRI come from a variety of fields, including engineering (electrical, mechanical, industrial, and design), computer science (human-computer interaction, artificial intelligence, robotics, natural language understanding, and computer vision), social sciences (psychology, cognitive science, communications, anthropology, and human factors), and humanities (ethics and philosophy).

Robots are poised to fill a growing number of roles in today's society, from factory automation to service applications to medical care and entertainment. While robots were initially used in repetitive tasks where all human direction is given a priori, they are becoming involved in increasingly more complex and less structured tasks and activities, including interaction with people required to complete those tasks. This complexity has prompted the entirely new endeavor of Human-Robot Interaction (HRI), the study of how humans interact with robots, and how best to design and implement robot systems capable of accomplishing interactive tasks in human environments. The fundamental goal of HRI is to develop the principles and algorithms for robot systems that make them capable of direct, safe and effective interaction with humans. Many facets of HRI research relate to and draw from insights and principles from psychology, communication, anthropology, philosophy, and ethics, making HRI an inherently interdisciplinary endeavor.

2.2.4 Mike Beiter, Brian coltin, Somchaya Liemhetcharat - published a paper titled as "An Introduction to Robotics with Nao". This work presents the preliminary results of an ongoing project which aims to use humanoid robots as sign language tutors.

2.2.5 B. Mazor, B.L. Zeigler -published a paper titled as "The design of speech-interactive dialog for transaction-automation systems" The use of automatic speech recognition (ASR) technology to automate telephone-service transactions provides opportunities for significant operational cost savings, as well as for improvements in the quality of customer service. At GTE, our focus has been on developing speech-interactive systems, named OASIS, for service-order transactions. In this paper, we discuss our methodology for developing dialog, describe the dialog developed for our service-dis- connect application, and present data from a field trial of the disconnect system. Our design methodology addresses development of a transaction model, specification of dialog flow, design of language structures, and construction of system speech. In addition, following our approach, dialog flow and recognition vocabularies are developed in the context of the system recognition and interpretation capabilities. Primary attributes of our methodology include the following: structured representation of conversational transactions as a progressive flow of information elements; query language that follows the style of discourse to elicit predictable responses; use of recognition outcomes and corresponding system actions to define dialog flow; and generation of scalable solutions. To evaluate the effective- ness of our approach, we present results from the service-disconnect field trial. In general, users responded cooperatively, adhered to the structured interaction, rarely anticipated future information, and provided on-target answers to system queries.

2.3 Summary

From the above survey the robot is not only controlled by android applications or any other things they were given intelligence to learn and teach those actions again. The robot consists of sensors that could actually detect human actions through sensors. The robot also consists of motors, camera microphone and speakers. The robot will be able to do those actions again once it has finished learning.

CHAPTER 3

PROPOSED SYSTEM

3.1 FLOW DIAGRAM

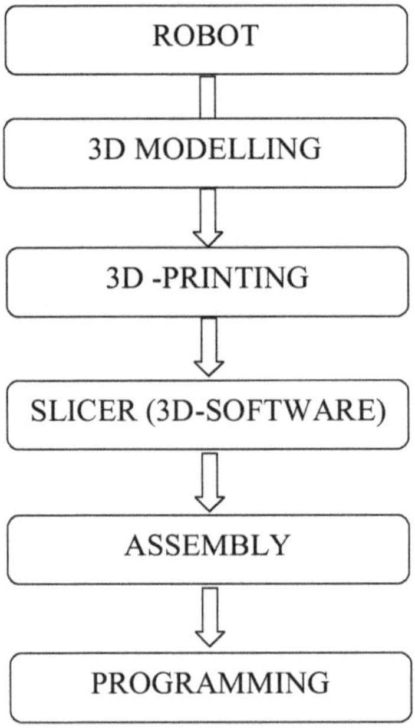

FIG 3.1 PROCESS FLOW

3.2 TECHNOLOGY

3.2.1 3D PRINTING

3D printing or additive manufacturing is a process of making three dimensional solid objects from a digital file. The creation of a 3D printed object is achieved using additive processes. In an additive process an object is created by laying down successive layers of material until the object is created. Each of these layers can be seen as a thinly sliced horizontal cross-section of the eventual object. Stereo Lithography (STL) is one of the most common file types that 3D printers can read. Thus, unlike material removed from a stock in the conventional machining process, 3D printing or AM builds a three-dimensional object from computer-aided design (CAD) model or AMF file by successively adding material layer by layer.3D printing is the opposite of subtract manufacturing which is cutting out / hollowing out a piece of metal or plastic with for instance a milling machine. The term "3D printing" originally referred to a process that deposits a binder material onto a powder bed with ink-jet printer heads layer by layer.3D printing enables you to produce complex (functional) shapes using less material than traditional manufacturing methods.

Applications include rapid prototyping, architectural scale models 3D printed prosthetics and movie props. Other examples of 3D printing would include reconstructing fossils in paleontology, replicating ancient artifacts in archaeology, reconstructing bones and body parts in forensic pathology and reconstructing heavily damaged evidence acquired from crime scene investigations.

3.2.2 PYTHON

Python is a high-level, interpreted, interactive and object-oriented scripting language. Python is designed to be highly readable. It uses English keywords frequently where as other languages use punctuation, and it has fewer syntactical constructions than other languages.

Python is interpreted, interactive, and its object oriented. Python is processed at runtime by the interpreter. You do not need to compile your program before executing it. This is similar to PERL and PHP. You can actually sit at a Python prompt and interact with the interpreter directly to write your programs. Python supports Object-Oriented style or technique of programming that encapsulates code within objects. Python has amazing libraries that really help you save time and cut down on the initial development cycle. Python has an excellent selection of libraries, from NumPy and SciPy for scientific computing to Django for web development.

There are even a few libraries with a more specific focus, like scikit-learn for machine learning applications and nltk for natural language processing.

The large standard library included in this language supports several programming tasks like searching text along with systematic expressions, linking to web servers, modifying and reading files. And it is a simple to access language, which makes it easy to achieve the program working. This feature makes Python as the ideal programming language for prototype development as well as other ad-hoc programming functions. There is no need to compromise maintainability.

3.2.3 AMAZON IMAGE AND OBJECT RECOGNITION

Amazon Rekognition makes it easy to add image and video analysis to your applications. You just provide an image or video to the Rekognition API, and the service can identify the objects, people, text, scenes, and activities, as well as detect any inappropriate content.

FIG 3.2 IMAGE RECOGNITON

Amazon Rekognition also provides highly accurate facial analysis and facial recognition. You can detect, analyze, and compare faces for a wide variety of user verification, cataloging, people counting, and public safety use cases.

Amazon Rekognition is based on the same proven, highly scalable, deep learning technology developed by Amazon's computer vision scientists to analyze billions of images and videos daily and requires no machine learning expertise to use. Amazon Rekognition is a simple and easy to use API that can quickly analyze any image or video file stored in Amazon S3. Amazon Rekognition is always learning from new data, and we are continually adding new labels and facial recognition features to the service.

Amazon Rekognition makes it easy to add visual analysis features to your application with easy to use APIs that don't require any machine learning expertise. Rekognition's APIs let you request range of features including object detection, facial recognition, and person tracking. With Amazon Rekognition, you can identify thousands of objects (e.g. bike, telephone, building) and scenes (e.g. parking lot, beach, city). When analyzing video, you can also identify specific activities happening in the frame, such as "delivering a package" or "playing soccer".

Rekognition's fast and accurate search capability allows you to identify a person in a photo or video using your private repository of face images. You can analyze the attributes of faces in images and videos to determine things like happiness, age range, eyes open, glasses, facial hair, etc. In video, you can also measure how these things change over time, such as constructing a timeline of the emotions of an actor.

You can quickly identify well known people in your video and image libraries to catalog footage and photos for marketing, advertising, and media industry use cases.

Specifically built to work with real world images, Rekognition can detect and recognize text from images, such as street names, captions, product names, and license plates.

Amazon Rekognition provides two API sets; Recognition Image for analyzing images and Rekognition Video for analyzing videos.

Both API perform detection and recognition analysis of images and videos to provide insights you can use in your applications. For example, you could use Rekognition Image to enhance the customer experience for a photo management application. When a customer uploads a photo, your application can use Rekognition Image to detect real-world objects or faces in the image. After your application stores the information returned from Rekognition Image, the user could then query their photo collection for photos with a specific object or face. Deeper querying is possible. For example, the user could query for faces that are smiling or query for faces that are a certain age.

For video applications, you could use Rekognition Video to create a surveillance application. Rekognition Video can track where a person is detected throughout a stored video. Alternatively, you can use Rekognition Video to search a streaming video for persons whose facial descriptions match facial descriptions already stored by Amazon Rekognition.

The Amazon Rekognition API makes deep learning image analysis easy to use. For example, Recognize Celebrities returns information for up to 100 celebrities detected in an image. This includes information about where celebrity faces are detected on the image and where to get further information about the celebrity.

The following information covers the types of analysis that Amazon Rekognition provides and an overview of Rekognition Image and Rekognition Video operations. Also covered is the difference between non-storage and storage operations.

Amazon Rekognition provides two API sets. Rekognition Image for analyzing images and Rekognition Video for analyzing stored and streaming videos. The following topic gives a brief overview of each API set.

The Rekognition Image and Rekognition Video API can detect or recognize a variety of entities such as faces or objects. For information about the types of recognition and detection that are supported, see Types of Detection and Recognition.

3.2.4 SPEECH RECOGNITION

Google Cloud Speech-to-Text enables developers to convert audio to text by applying powerful neural network models in an easy to use API. The API recognizes 120 languages and variants, to support your global user base. You can enable voice command-and-control, transcribe audio from call centers, and more. It can process real-time streaming or pre-recorded audio, using Google's machine learning technology.

Its powered by machine learning and Apply the most advanced deep learning neural network algorithms to audio for speech recognition with unparalleled accuracy. Cloud Speech-to-Text's accuracy improves over time as Google improves the internal speech recognition technology used by Google products.

Speech recognition works using algorithms through acoustic and language modeling. Acoustic modeling represents the relationship between linguistic units of speech and audio signals; language modeling matches sounds with word sequences to help distinguish between words that sound similar.

Often, hidden Markov models are used as well to recognize temporal patterns in speech to improve accuracy within the system.

Modern general-purpose speech recognition systems are based on Hidden Markov Models. These are statistical models that output a sequence of symbols or quantities. HMMs are used in speech recognition because a speech signal can be viewed as a piece wise stationary signal or a short-time stationary signal.

3.2.5 NETFABB

"Netfabb Studio is a software tailored for additive fabrication, rapid prototyping and 3D printing. It includes the ability of viewing, editing, repairing and analyzing stl-files or slice-based files in various formats.". Netfabb Studio is an STL editor, a must have tool for 3D printing, e.g with the Rap-Man, the fabbster or the Felix. There are other alternatives. Most slicer and control software can do some repairs. Mesh-mixer (from Autodesk who now also owns Netfabb) is a good alternative. Mesh lab allows difficult mesh repairs, but is difficult to use.

3.2.6 MESHLAB

The purpose of the Mesh-mixer is to convert a wide variety of 3D File Formats to Ogre Meshes via the Open Asset Import Library. It is useful for importing meshes exported from other 3D packages into Ogre: allows a large variety of post processing operations on the mesh, and supports all the formats that the Open Asset Import Library supports.

Mesh-lab is an advanced 3D mesh processing software system that is oriented to the management and processing of unstructured large meshes and provides a set of tools for editing, cleaning, healing, inspecting, rendering, and converting these kinds of meshes. Mesh-lab is free and open-source software, subject to the requirements of the GNU General Public License (GPL), version 2 or later, and is used as both a complete package and a library powering other software. It is well known in the more technical fields of 3D development and data handling.

CHAPTER 4

SYSTEM REQUIREMENTS

4.1 HARDWARE REQUIREMENTS

4.1.1 ARDUINO MEGA

The Arduino Mega 2560 is a microcontroller board based on the ATmega2560. It has 54 digital input/output pins (of which 15 can be used as PWM outputs), 16 analog inputs, 4 UARTs (hardware serial ports), a 16 MHz crystal oscillator, a USB connection, a power jack, an ICSP header, and a reset button. It contains everything needed to support the microcontroller; simply connecting it to a computer with a USB cable or power it with a AC-to-DC adapter or battery to get started. The Mega 2560 board is compatible with most shields designed for the Uno.

The MEGA 2560 is designed for more complex projects. With 54 digital I/O pins, 16 analog inputs and a larger space for sketch it is the recommended board for 3D printers and robotics projects.

FIG 4.1 ARDUINO MEGA

4.1.2 SERVO MOTOR (ACTUATORS):

Servo motors have been around for a long time and are utilized in many applications. They are small in size but pack a big punch and are very energy-efficient. These features allow them to be used to operate remote-controlled or radio-controlled toy cars, robots and airplanes. Servo motors are also used in industrial applications, robotics, in-line manufacturing, pharmaceutics and food services. To fully understand how the servo works, you need to take a look under the hood. Inside there is a pretty simple set-up: a small DC motor, potentiometer, and a control circuit. The motor is attached by gears to the control wheel. As the motor rotates, the potentiometer resistance changes, so the control circuit can precisely regulate how much movement there is and in which direction.

4.1.2.1 SERVO MOTOR SPECIFICATIONS:

Servo Motor Specification for fingers

1. Weight : 55g
2. Dimension : 40.7 × 19.7 × 42.9 mm

3. Operating Speed : 1) 20sec / 60 deg (4.8V no load) 2) 16sec / 60 deg (6.0V no load)
4. Stall Torque : 1) 10 kg-cm at 4.8V 2) 12 kg-cm at 6V
5. Operation Voltage : 4.8 - 7.2Volts
6. Gear Type: All Metal Gears
7. Dead band width: 5 μs
8. Stable and shock proof double ball bearing design
9. Temperature range: 0 ℃ – 55 ℃.
10. Control System : Analog
11. Operating Angle : 120degree
12. Required Pulse : 900us-2100us

FIG 4.2 SERVO MOTOR FOR FINGER

Servo motor specification for shoulder bicep and head

1. Weight : 200g
2. Dimension : 65.8X30.2X57.4mm
3. Operating Speed :   Stall Torque : 1) 48 kg-cm at 6V   2)   62 kg-cm at 7.4V
4. Operation Voltage : 6.0 – 7.4Volts
5. Gear Type: All Metal Gears
6. Dead band width: 4 μs
7. Stable and shock proof double ball bearing design
8. Temperature range: 0 °C – 55 °C.
9. Control System : Analog

FIG 4.3 SERVO MOTOR FOR SHOULDER BICEP AND HEAD

Servo motor specification for eye

1. Weight :8.5g
2. Dimension :32X11.5X24mm
3. Operating Speed : Stall Torque : 1) 1.5 kg-cm at 4.8V 2) 1.8 kg-cm at 6.0V
4. Operation Voltage :4.8 – 6.0Volts
5. Gear Type: All Metal Gears
6. Dead band width: 4 μs
7. Stable and shock proof double ball bearing design
8. Temperature range: 0 ºC – 55 ºC.
9. Control System :Analog

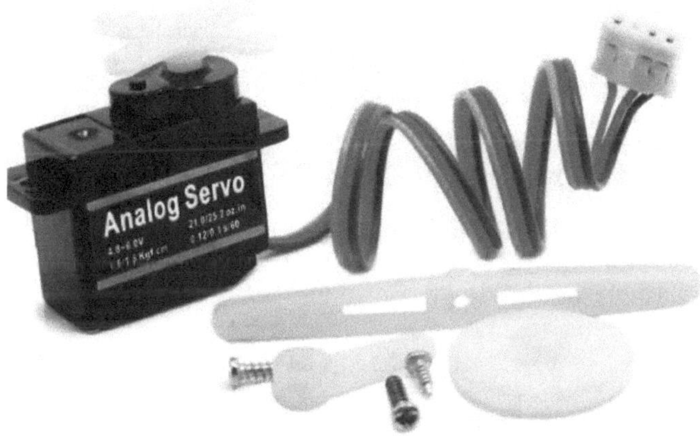

FIG 4.4 SERVO MOTOR FOR EYE MECHANISM

4.1.3 MEGA SENSOR SHIELD

It can connect to various modules like sensors, servos, relays, buttons, and potentiometers. Each functional module has buckled a port with VCC, GND and Output, which has a corresponding port on the Sensor Shield. Make connections easier, secure and more professional.

FIG 4.5 SENSOR SHIELD

4.1.4 CAMERA

Imaging Features Sensor VGA CMOS sensor technology Resolution • Motion Video: 640X480 pixel resolution* • Still Image: 1.3 megapixel interpolated* Field of View 55° diagonal field of view Imaging

Features • Digital pan, digital tilt, vertical tilt, and swivel pan, and 4x digital zoom** • Fixed focus from 0.4 m to 2.0 m • Automatic image adjustment with manual override

FIG 4.6 CAMERA

4.1.5 SWITCH MODE POWER SUPPLY

SMPS accepts input voltage of 220v and it produces an output of 5v dc. The output current produced is 10 amperes. It includes various applications such as cctv, led applications and vast field like cnc, robotics and DIY projects.

1.Input voltage: 220 v ac

2.Output voltage: 5v dc

3.Output current: 10 ampere

4. Application: cctv, led, cnc, robotics, DIY projects

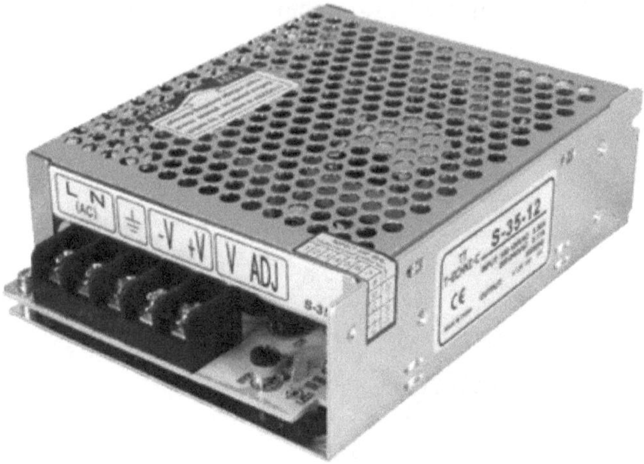

FIG 4.7 SWITCH MODE POWER SUPPLY

4.2 SOFTWARE REQUIREMENTS

4.2.1 SIMPLIFY 3D

Simplify 3D is a 3D printer slicing application. Where it reads a stereo lithographic file and slices into a g-code file where this g-codes are the geometrical codes and machine codes which generates the settings for the movements of the hot end of 3d-printer and sets the maximum and minimum temperature range of an hot end and heat-bed in a 3d printer, simplify3D gives the whole setting in a g-code file.

4.2.2 ZBRUSH

. It uses a proprietary "pixel" technology (see below) which stores lighting, color, material, and depth information for all objects on the screen. The main difference between Z Brush and more traditional modeling packages is that it is more akin to sculpting. Z Brush is used for creating high-resolution models (able to reach 40+ million polygons) for use in movies, games, and animations, by companies ranging from ILM to Electronic Arts. Z Brush uses dynamic levels of resolution to allow sculptors to make global or local changes to their models. Z Brush is most known for being able to sculpt medium to high frequency details that were traditionally painted in bump maps. The resulting mesh details can then be exported as normal maps to be used on a low poly version of that same model. This feature lets users work with complicated scenes without heavy processor overhead.

4.2.3 PYTHON 3.6

Python is developed under an OSI-approved open source license, making it freely usable and distributable, even for commercial use. The Python Package Index (PyPI) hosts thousands of third-party modules for Python. Both Python's standard library and the community-contributed modules allow for endless possibilities. Python is widely used in scientific and numeric computing and SciPy is a collection of packages for mathematics, science, and engineering, Pandas is a data analysis and modeling library including many more vast libraries.

4.2.4 AMAZON REKOGNITION

Amazon Rekognition makes it easy to add image and video analysis to your applications. You just provide an image or video to the Rekognition API, and the service can identify the objects, people, text, scenes, and activities, as well as detect any inappropriate content. Amazon Rekognition also provides highly accurate facial analysis and facial recognition. You can detect, analyze, and compare faces for a wide variety of user verification, cataloging, people counting, and public safety use cases.

Amazon Rekognition is based on the same proven, highly scalable, deep learning technology developed by Amazon's computer vision scientists to analyze billions of images and videos daily and requires no machine learning expertise to use. Amazon Rekognition is a simple and easy to use API that can quickly analyze any image or video file stored in Amazon S3. Amazon Rekognition is always learning from new data, and we are continually adding new labels and facial recognition features to the service.

CHAPTER 5

DESIGN AND PROCEDURE

5.1 SOLIDWORKS DESIGN

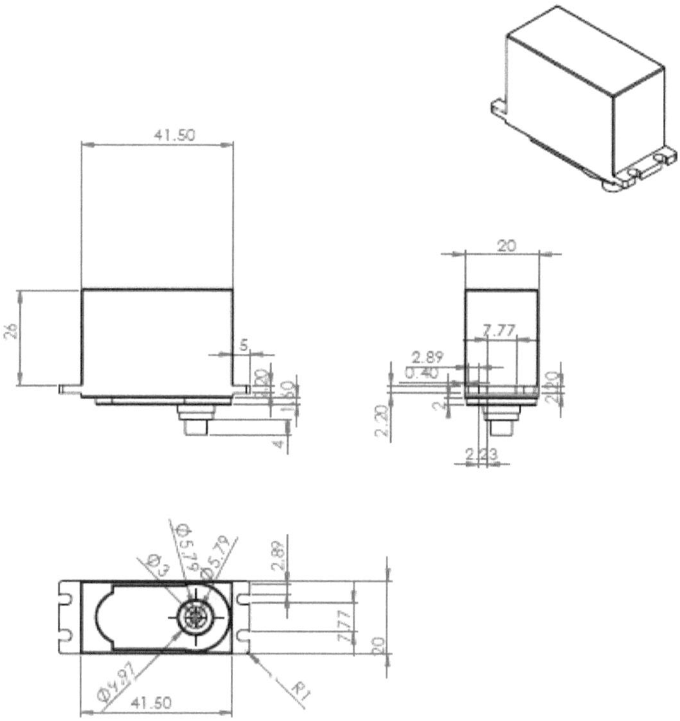

FIG 5.1 SERVO MOTOR

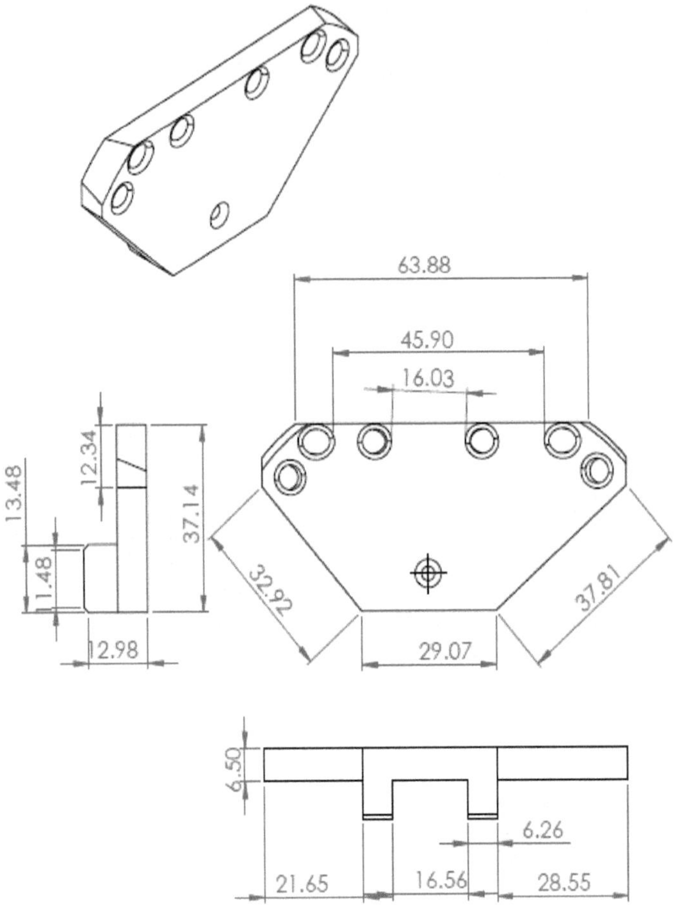

FIG 5.2 WIRE GUIDER

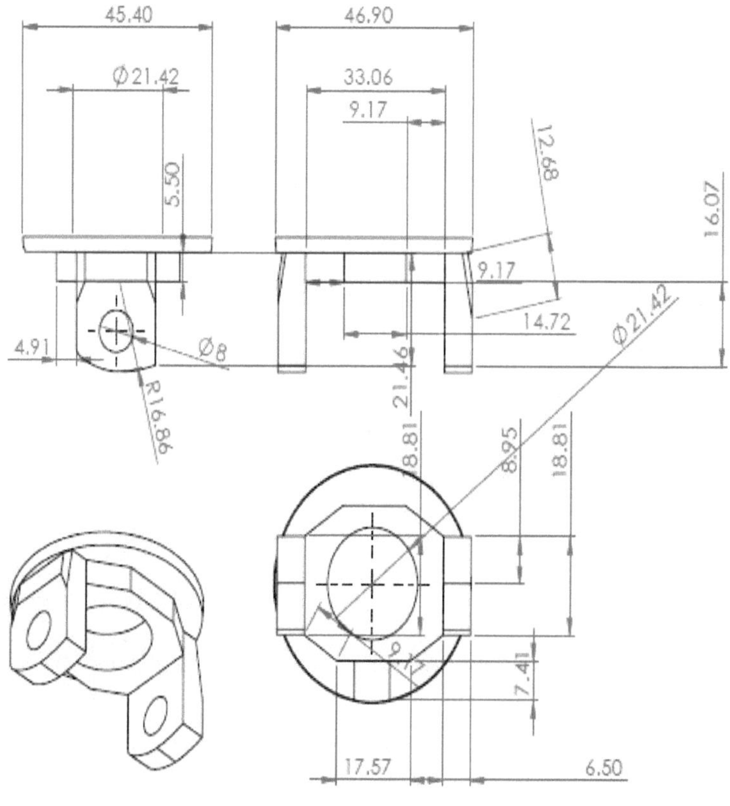

FIG 5.3 ROTATONAL WRIST

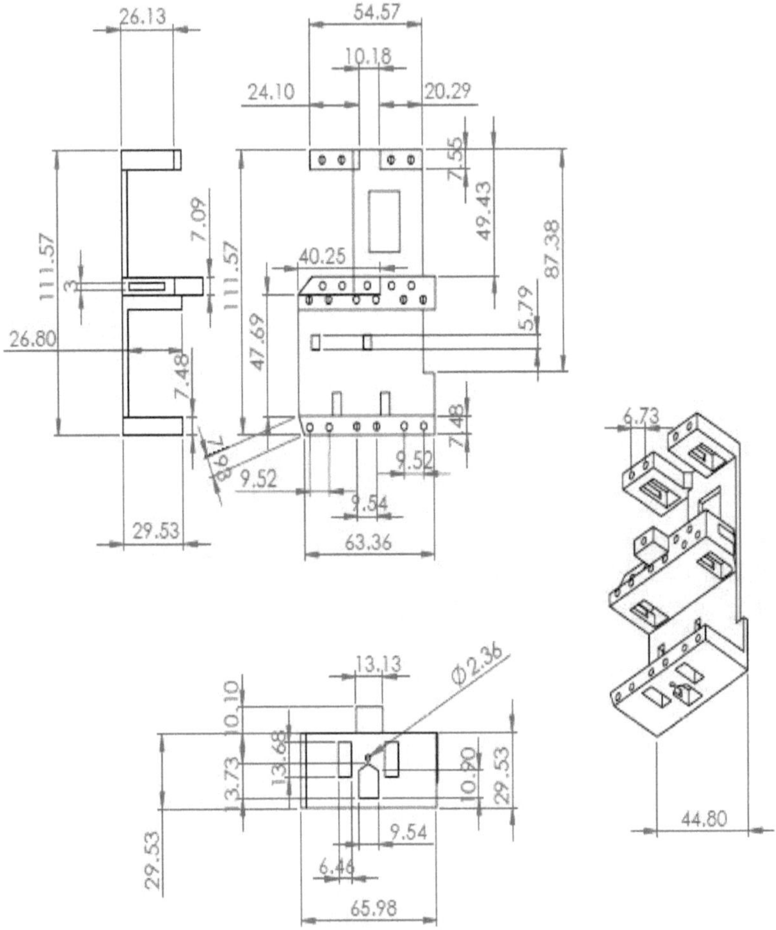

FIG 5.4 SERVOHOLDER

5.2 SIMPLIFY 3D

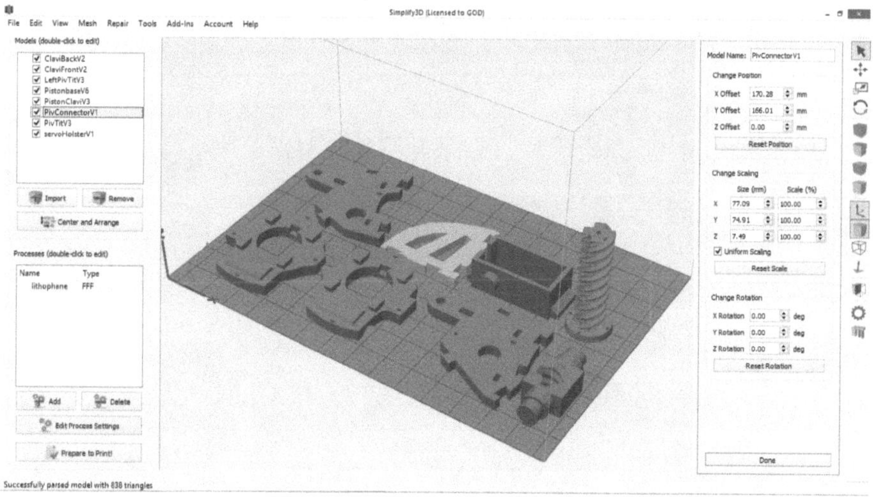

FIG 5.5 STL PLACEMENT

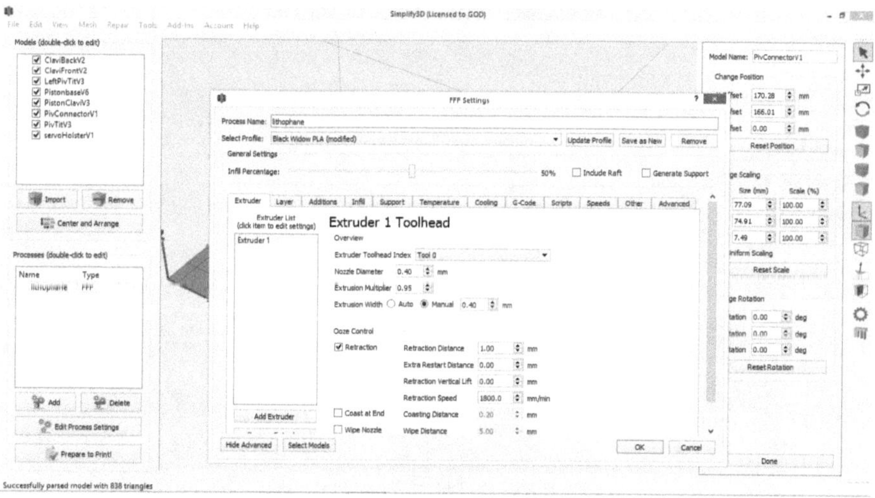

FIG 5.6 EXTRUDER SETTINGS

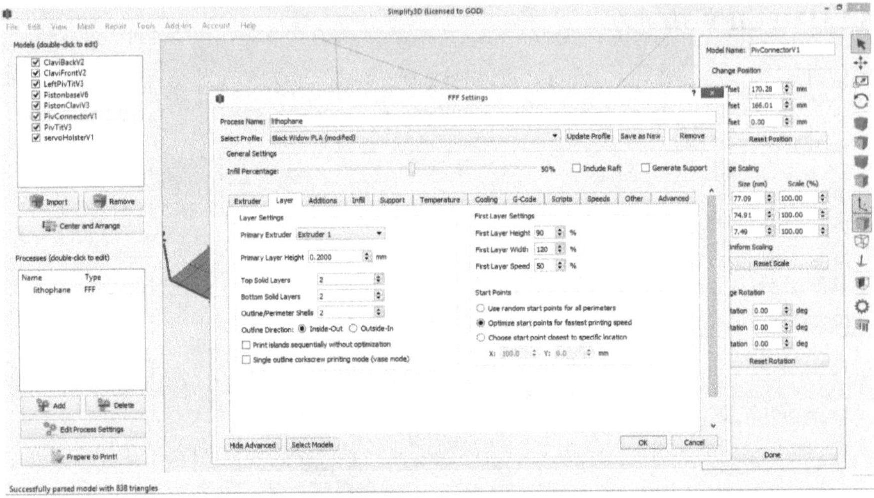

FIG 5.7 LAYER SETTINGS

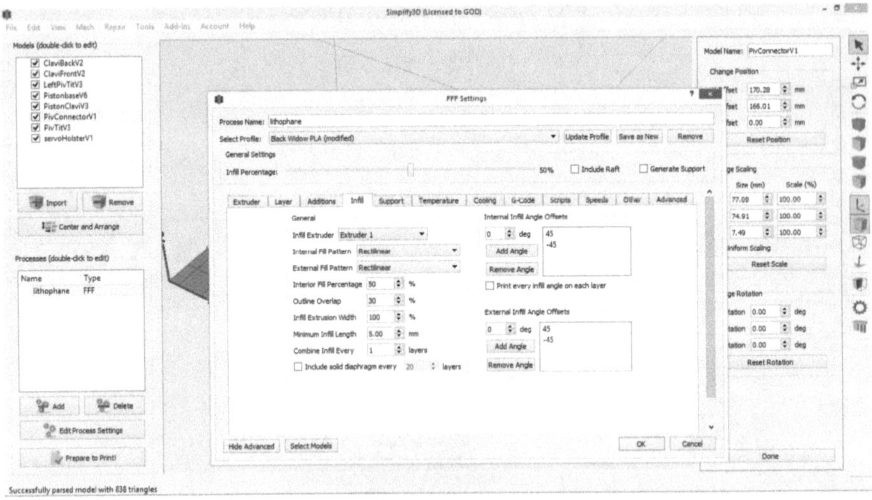

FIG 5.8 INFILL SETTINGS

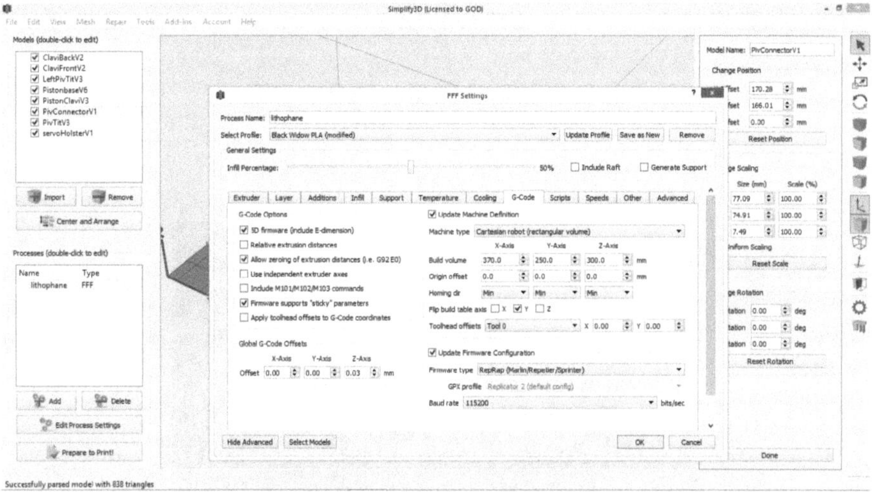
Simplify3D (Licensed to GOD)
File Edit View Mesh Repair Tools Add-ins Account Help
Models (double-click to edit)
ClaviBackV2
ClaviFrontV2
LeftPivTitV3
PistonbaseV6
PistonClaviV3
PivConnectorV1
PivTitV3
servoHolsterV1
Import
Remove
Center and Arrange
Processes (double-click to edit)
Name
Type
lithophane
FFF
Add
Delete
Edit Process Settings
Prepare to Print!
FFF Settings
Process Name: lithophane
Select Profile: Black Widow PLA (modified)
Update Profile
Save as New
Remove
General Settings
Infill Percentage:
50%
Include Raft
Generate Support
Extruder
Layer
Additions
Infill
Support
Temperature
Cooling
G-Code
Scripts
Speeds
Other
Advanced
G-Code Options
5D firmware (include E-dimension)
Relative extrusion distances
Allow zeroing of extrusion distances (i.e. G92 E0)
Use independent extruder axes
Include M101/M102/M103 commands
Firmware supports "sticky" parameters
Apply toolhead offsets to G-Code coordinates
Global G-Code Offsets
X-Axis
Y-Axis
Z-Axis
Offset
0.00
0.00
0.03
mm
Update Machine Definition
Machine type
Cartesian robot (rectangular volume)
Build volume
370.0
250.0
300.0
mm
Origin offset
0.0
0.0
0.0
mm
Homing dir
Min
Min
Min
Flip build table axis
X
Y
Z
Toolhead offsets
Tool 0
X
0.00
Y
0.00
Update Firmware Configuration
Firmware type
RepRap (Marlin/Repetier/Sprinter)
GPX profile
Baud rate
115200
bits/sec
Hide Advanced
Select Models
OK
Cancel
Model Name:
PivConnectorV1
Change Position
170.28
166.01
0.00
mm
Reset Position
Size (mm)
Scale (%)
77.09
74.91
7.49
100.00
Reset Scale
Reset Rotation
0.00
deg
Done
Successfully parsed model with 838 triangles

FIG 5.9 G-CODE

CHAPTER 6

IMPLEMENTATION

FIG 6.1 3D PRINTED COMPONENTS

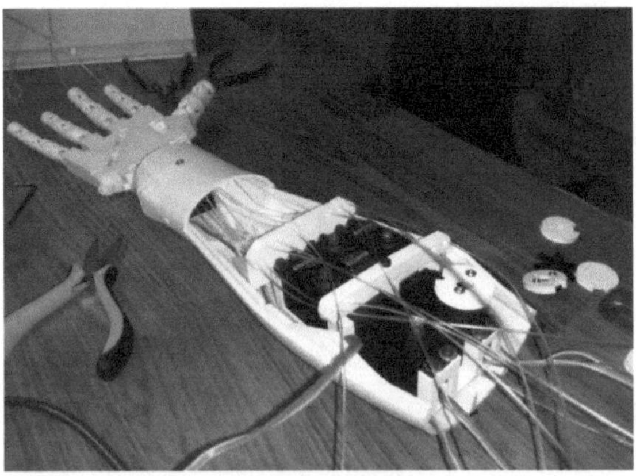

FIG 6.2 ARM WITH MOTORS ASSEMBLY

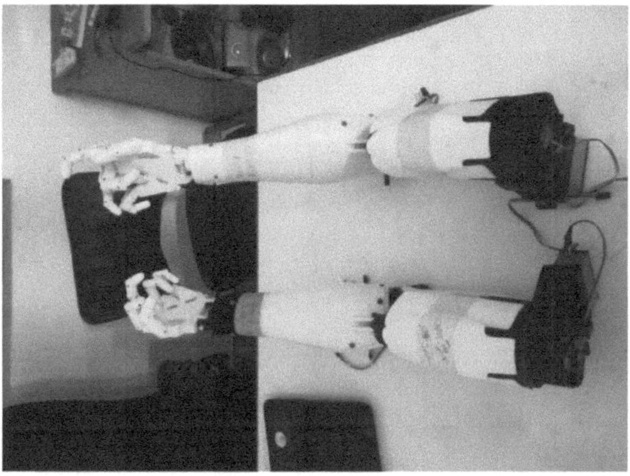

FIG 6.3 ROBOTIC ARM

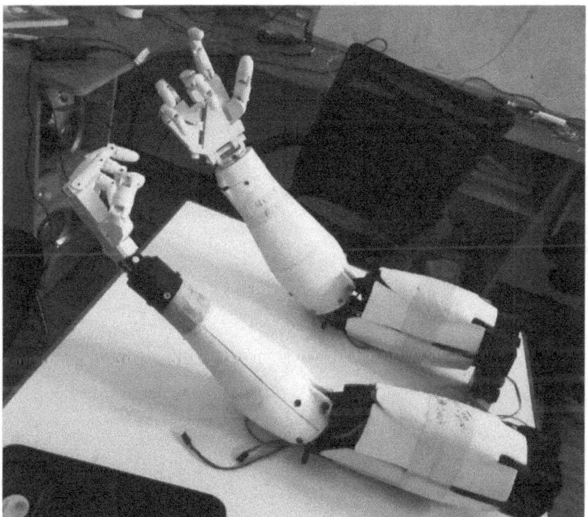

FIG 6.4 LR HAND

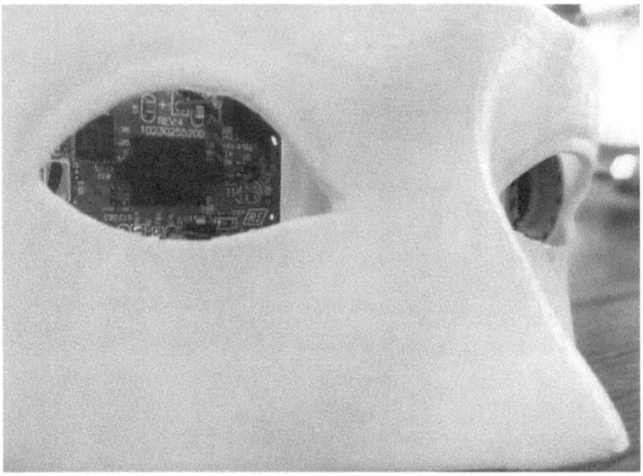

FIG 6.5 3D PRINTED EYE

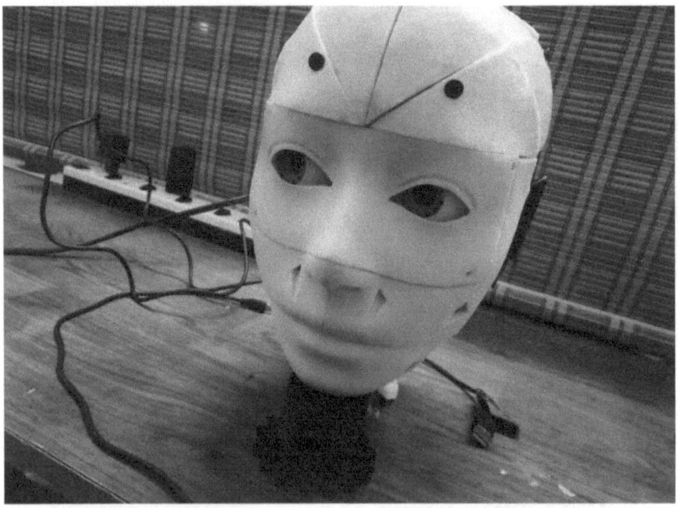

FIG 6.6 3D PRINTED HEAD

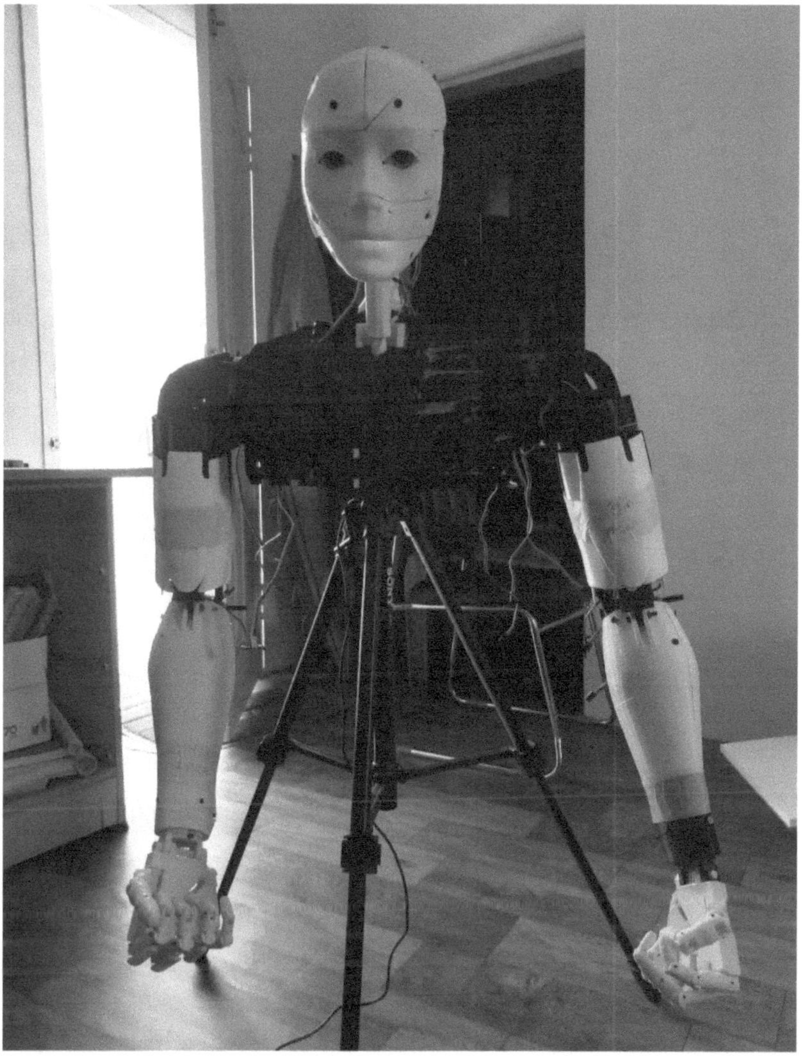

FIG 6.7 ROBOT

CHAPTER 7

CONCLUSION

This study has been carried out as part of an ongoing research, with the aim of helping to teach sign language to hearing impaired children by generating interaction based games between a humanoid robot and children. In this study, an interaction game is proposed with a set of 15 signs from Turkish Sign Language with similar gestures to test the success rate of the test participants without any previous knowledge of sign language.

CHAPTER 8

FUTURE ENHANCEMENT

The future of the project is to focus on optimizing the current design to create a 2-arm setup. covering facial expressions and the implementation of an expressive face to the design... Also, by adding 2 leg setup it can be transformed into a complete walkable humanoid robot. Where then it's a complete of a human it can speak, walk, play, pick loads and it covers most of the work humans can do. The main purpose of doing such a humanoid robot is to have AI powered human assist where it can handle problems faced by humans. It can be employed for jobs which are dangerous and which cannot be handled by humans. With the help of Neural networks and deep learning human activities can be easily mimicked by robots. Various motors and efficient and high cost high efficient motor can also be installed depending upon the requirement.

APPENDIX:

Program for humanoid robot hand:

```
#include<Servo.h>
Servo thumb_fingerservo;
Servo index_fingerservo;
Servo middle_fingerservo;
Servo ring_fingerservo;
Servo little_fingerservo;
void setup() {
 thumb_fingerservo.attach(0);
 index_fingerservo.attach(1);
 middle_fingerservo.attach(2);
 ring_fingerservo.attach(3);
 little_fingerservo.attach(4);
}

void loop() {
 hand_open();
 delay(1500);
 hand_close();
 delay(1500);
 thumb_open();
 delay(1500);
 hand_close();
 delay(1500);
 index_open();
 delay(1500);
 hand_close();
 delay(1500);
 middle_open();
 delay(1500);
 hand_close();
 delay(1500);
 ring_open();
 delay(1500);
 hand_close();
 delay(1500);
 little_open();
 delay(1500);
 hand_close();
 delay(2500);
 special_motion();
 delay(1500);
 teach();
 delay(1500);
 hand_close();
 delay(2500);
}

void hand_open()
{
 thumb_fingerservo.write(170);
 index_fingerservo.write(0);
 middle_fingerservo.write(0);
 ring_fingerservo.write(0);
 little_fingerservo.write(100);
}

void hand_close()
{index_fingerservo.write(180);
 middle_fingerservo.write(170);
 ring_fingerservo.write(170);
 little_fingerservo.write(0);
 thumb_fingerservo.write(25);

}

void thumb_open()
{
 thumb_fingerservo.write(180);
 index_fingerservo.write(180);
 middle_fingerservo.write(180);
 ring_fingerservo.write(180);
 little_fingerservo.write(15);
```

```
}

void index_open()
{
 thumb_fingerservo.write(30);
 index_fingerservo.write(0);
 middle_fingerservo.write(180);
 ring_fingerservo.write(180);
 little_fingerservo.write(15);
}

void middle_open(){
 thumb_fingerservo.write(0);
 index_fingerservo.write(180);
 middle_fingerservo.write(0);
 ring_fingerservo.write(180);
 little_fingerservo.write(15);
}
void ring_open(){
 thumb_fingerservo.write(0);
 index_fingerservo.write(180);
 middle_fingerservo.write(180);
 ring_fingerservo.write(0);
 little_fingerservo.write(15);
}

void little_open(){
 thumb_fingerservo.write(0);
 index_fingerservo.write(180);
 middle_fingerservo.write(180);
 ring_fingerservo.write(180);
 little_fingerservo.write(150);
}

void special_motion()
{
 thumb_fingerservo.write(180);
 delay(1000);
 index_fingerservo.write(0);
 delay(1000);
 middle_fingerservo.write(0);
 delay(1000);
 ring_fingerservo.write(0);
 delay(1000);
 little_fingerservo.write(150);
 delay(1000);
}

void teach(){
 index_fingerservo.write(0);
 delay(1000);
 index_fingerservo.write(180);
 delay(1000);
 index_fingerservo.write(0);
 middle_fingerservo.write(0);
 delay(1000);
 index_fingerservo.write(170);
 middle_fingerservo.write(170);
 delay(1000);
 index_fingerservo.write(0);
 middle_fingerservo.write(0);
 ring_fingerservo.write(0);
 delay(1000);
 index_fingerservo.write(170);
 middle_fingerservo.write(170);
 ring_fingerservo.write(170);
 delay(1000);
 index_fingerservo.write(0);
 middle_fingerservo.write(0);
 ring_fingerservo.write(0);
 little_fingerservo.write(100);
 delay(1000);
 index_fingerservo.write(170);
 middle_fingerservo.write(170);
 ring_fingerservo.write(170);
 little_fingerservo.write(0);
 delay(1000);
 thumb_fingerservo.write(180);
```

```
 index_fingerservo.write(0);
 middle_fingerservo.write(0);
 ring_fingerservo.write(0);
 little_fingerservo.write(100);
 delay(1000);
 index_fingerservo.write(30);
 middle_fingerservo.write(170);
 ring_fingerservo.write(170);
 little_fingerservo.write(100);
 thumb_fingerservo.write(25);
 delay(2000);
}
```

Program for humanoid robot head test:

```
from pymata_aio.pymata3 import PyMata3
servo=PyMata3()
head_u_d=4
head_l_r=5
eye_lr=2
eye_ud=26
mouth=3
def setup():
   servo.servo_config(head_l_r)
   servo.servo_config(head_u_d)
   servo.servo_config(eye_lr)
   servo.servo_config(eye_ud)
   servo.servo_config(mouth)

def loop():
   head_initial_position()
   eye_initial()
   head_motion()
   eye_initial()
   head_initial_position()
   headud()
   head_initial_position()
   eye_initial()

def head_initial_position():
   print("initialised head normal position")
   servo.analog_write(head_u_d,80)
   servo.sleep(0.050)
   servo.analog_write(head_l_r,100)
   servo.analog_write(mouth,100)
   servo.sleep(1)

def head_motion():
   print("initialised hesd motion")
   servo.analog_write(head_u_d,20)
   servo.sleep(1)
   k=100
   m=50
   for i in range(70,130):
      servo.analog_write(head_l_r,i)
      servo.sleep(0.030)
      servo.analog_write(eye_lr,k)
      servo.sleep(.050)
      if k>50:
         k=k-1

   for j in range(130,70,-1):
      servo.analog_write(head_l_r,j)
      servo.sleep(0.030)
      servo.analog_write(eye_lr,m)
      servo.sleep(.050)
      if m<100:
         m=m+1

def eye_initial():
   print("initialised eye normal postion")
   servo.analog_write(eye_lr,70)
   servo.sleep(.050)
   servo.analog_write(eye_ud,100)
   servo.sleep(1)
```

```
def eye_motion():
   print("initialised eye motion")
   for k in range(50,110):
      servo.analog_write(eye_lr,k)
      servo.sleep(.050)
   for m in range(50,110,-1):
      servo.analog_write(eye_lr,m)
      servo.sleep(.050)

def headud():
   print("initialised head up and down")
   k=110
   m=60
   for i in range(20,90,+2):
      servo.analog_write(head_u_d,i)
      servo.sleep(0.010)
      servo.analog_write(eye_ud,k)
      servo.sleep(0.050)
      if k>60:
         k=k-2
         servo.sleep(0.020)
   for j in range(90,20,-2):
      servo.analog_write(head_u_d,j)
      servo.sleep(0.010)
      servo.analog_write(eye_ud,m)
      servo.sleep(0.050)
      if k<100:
         m=m+1
         servo.sleep(0.020)
if __name__=="__main__":
   setup()
   while True:
      loop()
      break
```

REFERENCES

[1] Kose-Bagci, H., E. Ferrari, K. Dautenhahn, D. S. Syrdal, and C.L. Nehaniv, "Effects of Embodiment and Gestures on Social Interaction in Drumming Games with a Humanoid Robot",Special issue on Robot and Human Interactive Communication, Advanced Robotics vol.24, no.14, 2009.

[2] Kose-Bagci, H, .K. Dautenhahn, D. S. Syrdal, and C. L. Nehaniv,"Effects of embodiment and gestures on social interaction in drumming games with a humanoid robot," Connection Science, vol. 22, no. 2, pp. 103– 134, 2010.

[3] Kose-Bagci, H., R. Yorganci , H. E. Algan, ""Evaluation of the Robot Sign Language Tutor using video-based studies", 5th European Conference on Mobile Robots (ECMR11), 7th-9th September,2011.

[4] Kose-Bagci, H., R. Yorganci , "Tale of a robot: Humanoid Robot Assisted Sign Language Tutoring", 11th IEEE-RAS International Conference on Humanoid Robots, Bled, Slovenia(HUMANOIDS 2011) , Oct. 26-28, 2011.

[5] Kose-Bagci, H., R. Yorganci , H. E. Algan, and D.S. Syrdal,"Evaluation of the Robot Assisted Sign Language Tutoring sing video-based studies", SORO special issue on "Measuring Human-Robot Interaction", 2011.

[6] Y. Sakagami, R. Watanabe, C. Aoyama, M. Shinichi, N. Higaki, K. Fujimura, "The intelligent ASIMO: System overview and integration", Proceedings of the IEEE International Conference on Intelligent Robots and Systems, October 2002.

[7] T. Sawada, T. Takagi, M. Fujita, "Behavior Selection and Motion Modulation in Emotionally Grounded Architecture for QRIO SDR-4X II", Proceedings of the IEEE International Conference on Intelligent Robots and Systems, vol. 3, pp. 2514-2519, October 2004.

[8] S. Lohmeier, K. Löffler, M. Gienger, H. Ulbrich, F. Pfeiffer, "Computer System and Control of Biped "Johnnie"", Proceedings of the IEEE International Conference on Robotics and Automation, vol. 4, pp. 4222-4227, April 2004.

[9] K. Kaneko, F. Kanehiro, S. Kajita, K. Yokoyama, K. Akachi, T. Kawasaki, S. Ota, T. Isozumi, "Design of Prototype Humanoid Robotics Platform for HRP", IEEE International Conference on Intelligent Robots and Systems, vol. 3, pp. 2431-2436, October 2002.

[10] K. Hirai, M. Hirose, Y. Haikawa, T. Takenaka, "The Development of Honda Humanoid Robot", Proceedings of IEEE International Conference on Robotics and Automation, vol. 2, pp. 1321-1326, May 1998.

[11] S. Kajita, K. Tani, "Study of Dynamic Biped Locomotion on Rugged Terrain —Theory and Basic Experiment-", ICAR Fifth International Conference on Advanced Robotics, vol. 1, pp. 741-746, June 1991.

[12] K. Kaehiro, K. Kaneko, K. Fujiwara, K. Yokoi, H. Hirukawa, "A Real Time Pattern Generator for Bipedal Walking", Proceedings of the IEEE International Conference on Robotics and Automation, vol. 1, pp. 31-37, May 2002.

[13] S. Kajita, F. Kahehiro, K. Kaneko, K. Fujiwara, K. Harada, K. Yokoi, H. Hirukawa, "Biped Walking Pattern Generation using Preview Control of the Zero-Moment-Point", Proceedings of IEEE International Conference on Robotics and Automation, vol. 2, pp. 1620-1626, September 2003.

Kolibri 360 GmbH
Ferdinand-Jühlke-Straße 7
99095 Erfurt, Deutschland
produktsicherheit@kolibri360.de

Druck:
CPI Druckdienstleistungen GmbH
im Auftrag der
Kolibri 360 GmbH
Ferdinand-Jühlke-Str. 7
99095 Erfurt